JUMBO COLORING BOOK

GRASSLANDS ARE WIDE, OPEN AREAS WITH LOTS OF GRASS. THEY GROW WHERE THERE'S SOME RAIN-NOT TOO MUCH FOR FORESTS AND NOT TOO LITTLE LIKE DESERTS.

Biographical Note
Grassland Ecosystems An Educational Coloring Book is a new work,
first published by Little Artist Studio in 2025.

International Standard Book Number
ISBN 979-8-9992504-3-8

www.littleartiststudio.org

EXPERIENCE THE RICH DIVERSITY OF THE WORLD'S GRASSLAND ECOSYSTEMS WITH OVER 85 BEAUTIFULLY DETAILED COLORING PAGES. THIS EDUCATIONAL COLORING BOOK FEATURES ICONIC GRASSLANDS FROM ACROSS THE GLOBE, DESIGNED TO INSPIRE CURIOSITY AND DEEPEN UNDERSTANDING OF THESE VITAL ENVIRONMENTS. PART OF LITTLE ARTIST STUDIO'S ACCLAIMED EDUCATIONAL SERIES, EACH FULL-PAGE ILLUSTRATION TELLS A CAPTIVATING STORY THAT SPARKS CREATIVITY AND LEARNING. WITH SINGLE-SIDED PAGES, ARTISTS OF ALL AGES CAN USE ANY COLORING MEDIUM AND EASILY DISPLAY THEIR FINISHED MASTERPIECES. PERFECT FOR NATURE LOVERS, EDUCATORS, AND CREATIVE MINDS EVERYWHERE.

EURASIAN STEPPES

THE EURASIAN STEPPE (SOMETIMES SPELLED "STEPPES") IS A HUGE AREA OF GRASSLANDS THAT STRETCHES ACROSS PARTS OF EUROPE AND ASIA.

EURASIAN STEPPES

MONGOLIAN
WILD HORSE

EURASIAN STEPPES

EURASIAN STEPPES

MONGOLIAN
GAZELLE

EURASIAN STEPPES

WILD BACTRIAN CAMELS LIVE IN DRY
AREAS LIKE DESERTS, MOUNTAINS, AND
GRASSLANDS IN CHINA AND MONGOLIA.
THEY ARE IN DANGER BECAUSE OF
HUNTING, LOSING THEIR HOMES, AND
SHARING SPACE WITH FARM ANIMALS.

EURASIAN STEPPES

STEPPES HAVE LOTS OF GRASSES, BUSHES, AND FLOWERS THAT GROW IN DRY AREAS. SOME COMMON PLANTS ARE WHEATGRASS, NEEDLE-AND-THREAD GRASS, SAGEBRUSH, AND GREASEWOOD.

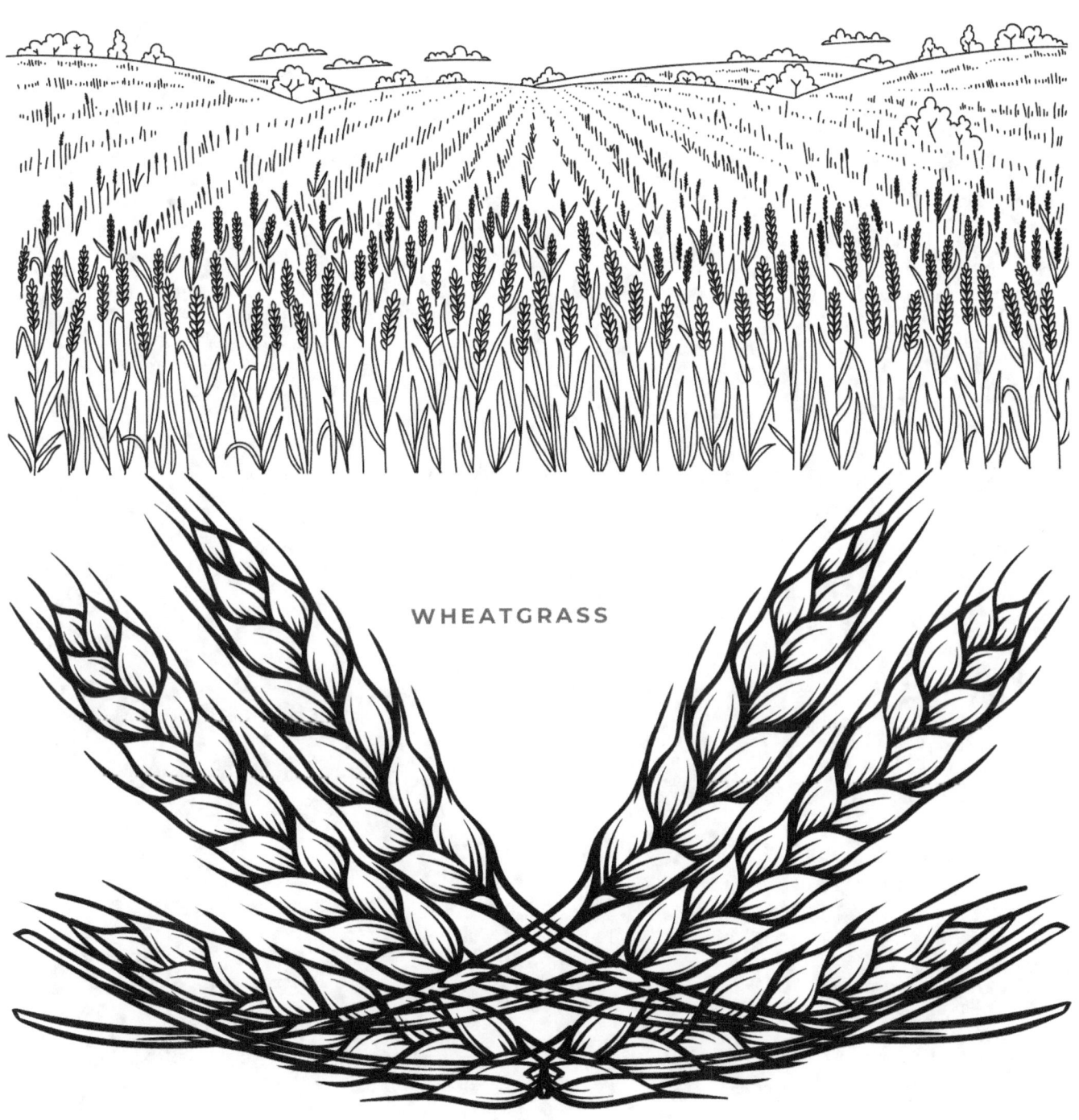

WHEATGRASS

EURASIAN STEPPES

SAGEBRUSH AND SAGE ARE NOT THE SAME.
SAGE IS A MINTY HERB, AND SAGEBRUSH IS
PART OF THE SUNFLOWER FAMILY. BUT
SAGEBRUSH FLOWERS DON'T LOOK LIKE
SUNFLOWERS AND ARE SPREAD BY THE
WIND, NOT BY INSECTS.

SAGE

EURASIAN STEPPES

SAGEBRUSH

EURASIAN STEPPES

SOME TYPES OF OWLS, LIKE THE SHORT-EARED OWL AND LITTLE OWL, LIVE IN STEPPE HABITATS.

LIFE ON THE STEPPES

MUCH OF THE EASTERN EURASIAN STEPPE IS COVERED BY DRYLAND GRASSES WHICH, WHILE CHALLENGING FOR GRAIN AGRICULTURE, CAN SUSTAIN LARGE MEAT AND DAIRY-PRODUCING HERD.

LIFE ON THE STEPPES

PEOPLE LIVING ON
THE STEPPES
HAVE RAISED
GRAZING ANIMALS
FOR THOUSANDS
OF YEARS. THESE
ANIMALS PRODUCE
MILK.

SUB-SAHARAN AFRICA

SUB-SAHARAN AFRICAN GRASSLANDS ARE BIG, OPEN PLACES WITH MOSTLY GRASS AND SOME TREES.

BLACK MAMBA

SUB-SAHARAN AFRICA

BAOBAB TREES, WITH THEIR MASSIVE TRUNKS AND DISTINCTIVE SHAPES, ARE A STRIKING FEATURE OF THE AFRICAN SAVANNA LANDSCAPE AND CAN BE FOUND IN MANY AREAS ACROSS IT.

SUB-SAHARAN AFRICA

BAOBABS ARE CALLED "THE TREE OF LIFE" BECAUSE THEY GIVE FOOD AND SHELTER TO MANY ANIMALS AND PLAY AN IMPORTANT ROLE IN THE ECOSYSTEM.

SUB-SAHARAN AFRICA

GALAGOS, ALSO CALLED BUSHBABIES, AND FRUIT BATS DRINK THE SWEET NECTAR FROM BAOBAB FLOWERS. WHILE THEY EAT, THEY HELP POLLINATE THE TREES.

SUB-SAHARAN AFRICA

WARTHOGS EAT PARTS OF THE BAOBAB TREE, ESPECIALLY THE FRUIT AND SEED PODS.

SUB-SAHARAN AFRICA

VENOMOUS SNAKES, SUCH AS THE BOOMSLANG, OFTEN MAKE THEIR HOMES IN BAOBAB TREE BRANCHES OR THE HOLLOW TRUNKS, USING THE TREE FOR SHELTER AND HUNTING.

SUB-SAHARAN AFRICA

BAOBAB TREES DON'T PROVIDE FOOD FOR VULTURES, BUT THEY OFFER SAFE PLACES TO NEST AND REST, OFTEN SERVING AS LOOKOUT POINTS. THESE TREES SUPPORT MANY ANIMALS IN THE SAVANNA.

SUB-SAHARAN AFRICA

SUB-SAHARAN AFRICA

MANY ANIMALS LIVE IN THE
GRASSLANDS. LIONS LIVE HERE TOO
BECAUSE THEY HAVE LOTS OF SPACE TO
HUNT.

NORTH AMERICAN PRAIRIES

NORTH AMERICAN PRAIRIES ARE BIG, GRASSY
AREAS WITH FEW TREES. THEY ARE SUNNY
AND FULL OF PLANTS LIKE GRASS AND
WILDFLOWERS. MANY ANIMALS LIVE THERE.

BISON

NORTH AMERICAN PRAIRIES

PRAIRIES ARE IMPORTANT BECAUSE THEY GIVE FOOD AND HOMES TO ANIMALS.

PRONGHORN

NORTH AMERICAN PRAIRIES

HERBIVORES, LIKE BISON AND PRAIRIE DOGS, PRIMARILY CONSUME GRASSES AND OTHER PRAIRIE PLANTS.

PRAIRIE DOG

NORTH AMERICAN PRAIRIES

OMNIVORES, SUCH AS RACCOONS AND
BEARS, EAT A MIX OF PLANTS, INSECTS,
AND SMALL ANIMALS.

RACCOON

NORTH AMERICAN PRAIRIES

CARNIVORES, INCLUDING COYOTES AND HAWKS, PREY ON HERBIVORES AND OTHER SMALLER ANIMALS.

CAYOTE

RABBIT

NORTH AMERICAN PRAIRIES

JACKRABBITS ARE ANIMALS WITH LONG
LEGS THAT LIVE IN THE BIG GRASSLANDS
OF NORTH AMERICA CALLED PRAIRIES.
THEY RUN FAST TO STAY SAFE AND EAT
GRASS AND PLANTS.

NORTH AMERICAN PRAIRIES

WHILE COMMONLY CALLED "JACKRABBITS," THESE ANIMALS ARE TECHNICALLY HARES - A TYPE OF RABBIT WITH LONGER LEGS AND EARS, AND BORN WITH FUR AND OPEN EYES.

HARE

RABBIT

NORTH AMERICAN PRAIRIES

GRASSLANDS STRETCH AS FAR AS THE EYE
CAN SEE. THE LAND IS MOSTLY FLAT OR
GENTLY ROLLING, WITH TALL GRASSES,
WILDFLOWERS, AND FEW TREES.

NORTH AMERICAN PRAIRIES

NORTH AMERICAN PRAIRIES

IN SPRING AND SUMMER, THE PRAIRIES ARE
FULL OF COLORFUL FLOWERS AND BUZZING
INSECTS-LIKE A GIANT OUTDOOR GARDEN!

NORTH AMERICAN PRAIRIES

FARM ANIMALS COMMONLY RAISED ON
THE NORTH AMERICAN PRAIRIES
INCLUDE CATTLE, PIGS, SHEEP,
CHICKENS, AND HORSES.

NORTH AMERICAN PRAIRIES

BURROWING OWLS ARE SMALL OWLS THAT
LIVE IN THE GROUND. THEY COME OUT
DURING THE DAY AND EAT BUGS AND
SMALL ANIMALS LIKE MICE.

ARGENTINE PAMPAS

THE ARGENTINE PAMPAS ARE WIDE WITH
RICH SOIL AND GOOD WEATHER, AND
GREAT FOR FARMING AND RAISING COWS.

ARGENTINE PAMPAS

LA PAMPA IS A BEAUTIFUL PLACE IN
ARGENTINA WITH GRASSLANDS, FORESTS,
CAVES, AND LAKES.

ARGENTINE PAMPAS

THE ARGENTINE PAMPAS ARE WIDE WITH
RICH SOIL AND GOOD WEATHER, AND
GREAT FOR FARMING AND RAISING COWS.

ARGENTINE PAMPAS

ARGENTINA IS FAMOUS FOR
CATTLERANCHING AND ITS HIGH-
QUALITY BEEF.

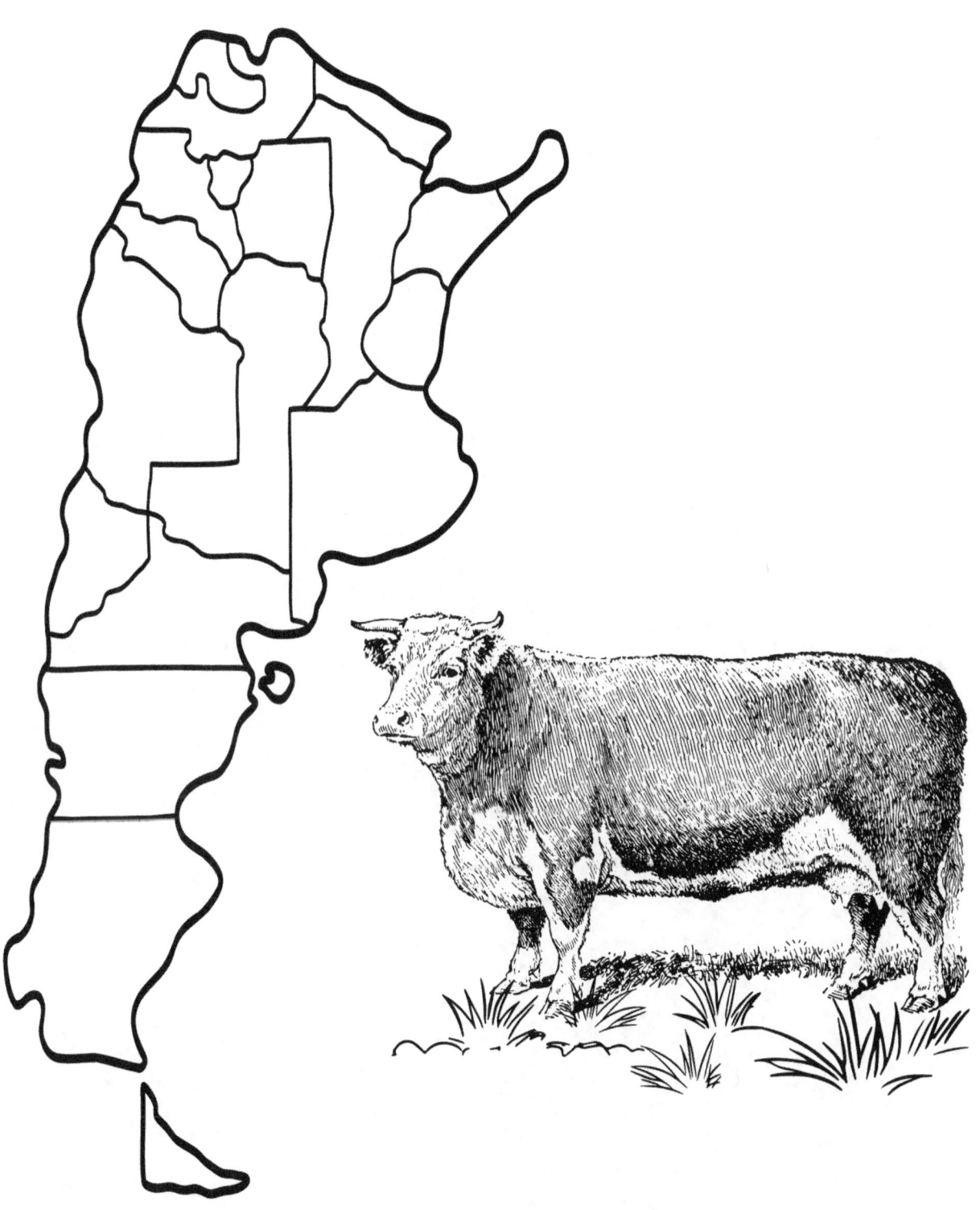

ARGENTINE PAMPAS

THE PAMPAS IS A BIG GRASSY AREA IN ARGENTINA. IN THIS REGION, PEOPLE RIDE HORSES AND TAKE CARE OF CATTLE AT RODEOS, WHICH ARE FUN EVENTS THAT SHOW THEIR SKILLS.

ARGENTINE PAMPAS

FARMERS IN THE PAMPAS GROW A LOT OF CORN BECAUSE THE LAND IS RICH AND THE WEATHER IS IDEAL. IT'S AN IMPORTANT CROP THAT HELPS FEED PEOPLE AROUND THE WORLD.

GORDITA

CORN COBS

TAMALE

CORNBREAD

POZOLE

POLENTA

ARGENTINE PAMPAS

SHEEP FARMING IS AN IMPORTANT ACTIVITY IN THE ARGENTINE PAMPAS. WIDE-OPEN SPACES ARE GREAT FOR RAISING ANIMALS LIKE SHEEP AND CATTLE.

NORTHERN AUSTRALIA

NORTHERN AUSTRALIA HAS BIG, OPEN AREAS
CALLED SAVANNAS. THESE GRASSLANDS HAVE
TALL GRASSES, SCATTERED TREES, AND A WET
AND DRY SEASON.

KANGAROO

JOEY

NORTHERN AUSTRALIA

LOTS OF ANIMALS LIKE KANGAROOS, EMUS, AND WALLABIES LIVE IN NORTHERN AUSTRALIA. IT'S WARM ALL YEAR, BUT RAINS MOSTLY COME DURING ONE PART OF THE YEAR (THE WET SEASON).

EMU

NORTHERN AUSTRALIA

FRESHWATER CROCODILES PLAY AN IMPORTANT ROLE IN BALANCING THE ECOSYSTEM, ACTING AS APEX PREDATORS THAT HELP REGULATE POPULATIONS OF FISH, FROGS, AND OTHER PREY SPECIES.

FRESHWATER CROCODILE

NORTHERN AUSTRALIA

IN THE DRY SEASON, ANIMALS LIKE KANGAROOS, BIRDS, AND FRESHWATER CROCODILES STAY NEAR THE WATER TO STAY COOL AND FIND FOOD.

FINCH

AUSTRALIAN MAGPIE

NORTHERN AUSTRALIA

FERAL HORSES, KNOWN AS
BRUMBIES, LIVE IN NORTHERN
AUSTRALIA.

NORTHERN AUSTRALIA

DINGOES ARE WILD DOGS
NATIVE TO AUSTRALIA AND ARE
CONSIDERED THE CONTINENT'S
TOP LAND PREDATOR.

FRILLED LIZARD

NORTHERN AUSTRALIA

THE EASTERN BROWN SNAKE, WITH FIXED
FRONT FANGS, IS CONSIDERED ONE OF THE
MOST DANGEROUS SNAKES IN THE WORLD.

NORTHERN AUSTRALIA

BRAHMAN CATTLE ARE STRONG COWS THAT LIVE IN HOT, DRY PARTS OF NORTHERN AUSTRALIA. THEY ARE GOOD AT STAYING COOL AND ARE RAISED BY FARMERS FOR MEAT.

NORTHERN AUSTRALIA

A LONG TIME AGO, PEOPLE BROUGHT CAMELS TO AUSTRALIA TO HELP CARRY THINGS IN THE DESERT. NOW, SOME OF THOSE CAMELS LIVE IN THE WILD AND ROAM THE HOT, DRY PARTS OF NORTHERN AUSTRALIA.

NORTHERN AUSTRALIA

THE BUTTON QUAIL IS A TINY, SHY BIRD THAT LIVES IN GRASSLANDS AND FORESTS IN AUSTRALIA. IT'S SMALL LIKE A BUTTON AND LOVES TO HIDE IN THE GRASS!

PLANTS AND ANIMALS

WALLABIES ARE SMALL ANIMALS THAT LOOK LIKE KANGAROOS AND LIVE IN NORTHERN AUSTRALIA. THEY HOP ON STRONG BACK LEGS AND EAT GRASS AND PLANTS.

PLANTS AND ANIMALS

KOALAS ARE MARSUPIALS, WHICH MEANS THEY ARE A TYPE OF MAMMAL THAT CARRIES AND RAISES THEIR BABIES IN A POUCH. THEY EAT EUCALYPTUS, ALSO CALLED GUM TREES. THEY ARE NOT BEARS, WE SWEAR!

KOALA

EUCALYPTUS

PLANTS AND ANIMALS

THE EVERLASTING DAISY, NATIVE TO AUSTRALIA,
AUSTRALIA, ARE FAMOUS FOR THEIR ABILITY TO
RETAIN THEIR COLOR AND SHAPE WHEN DRIED,
LASTING FOR A LONG TIME.

PLANTS AND ANIMALS

THE IMPALA IS A FAST ANTELOPE FROM AFRICA. IT HAS CURVED HORNS AND CAN JUMP REALLY HIGH TO ESCAPE DANGER!

PLANTS AND ANIMALS

THE AMERICAN BISON IS A LARGE,
FURRY ANIMAL FROM NORTH AMERICA.
IT HAS A BIG HEAD AND ONCE ROAMED
THE GRASSLANDS IN HUGE HERDS.

GRASSLANDS ECONOMICS

GRASSLAND ECONOMICS IS THE STUDY OF
HOW PEOPLE USE AND BENEFIT FROM
GRASSLANDS IN WAYS THAT AFFECT MONEY,
JOBS, AND NATURAL RESOURCES.

GRASSLANDS ECONOMICS

SUB-SAHARAN AFRICA'S GRASSLANDS PLAY A CRUCIAL ROLE IN THE ECONOMY THROUGH WILDLIFE TOURISM, PROVIDING EMPLOYMENT AND FUNDING FOR CONSERVATION WHILE PROMOTING SUSTAINABLE DEVELOPMENT.

GRASSLANDS ECONOMICS

MANY GRASSLANDS ARE USED TO GROW CROPS LIKE WHEAT OR CORN AND RAISE ANIMALS LIKE COWS AND SHEEP. THIS HELPS PRODUCE FOOD AND SUPPORTS JOBS FOR FARMERS AND RANCHERS.

GRASSLANDS ECONOMICS

GRASSLANDS HELP SUPPORT HEALTHY ECOSYSTEMS BY
STORING CARBON, CLEANING WATER, AND PREVENTING
SOIL EROSION, WHICH HAS LONG-TERM ECONOMIC VALUE
EVEN IF IT'S NOT ALWAYS DIRECTLY SEEN IN DOLLARS.

INVASIVE SPECIES

IN NORTHERN AUSTRALIA, PAMPAS GRASS IS CLASSIFIED AS AN ENVIRONMENTAL WEED AND REQUIRES CAREFUL MANAGEMENT BECAUSE IT IS HIGHLY INVASIVE AND CAN OUTCOMPETE NATIVE PLANT SPECIES.

INVASIVE SPECIES

LANTANA CAMARA IS A PLANT THAT
SPREADS TOO FAST IN AFRICAN
GRASSLANDS AND MAKES IT HARD FOR
LOCAL PLANTS AND ANIMALS TO LIVE.

INVASIVE SPECIES

FERAL PIGS ARE WILD PIGS THAT DON'T LIVE ON FARMS. THEY ORIGINALLY CAME FROM EUROPE AND ASIA BUT NOW LIVE IN MANY PARTS OF THE WORLD, INCLUDING SOME GRASSLANDS IN AFRICA.

INVASIVE SPECIES

AFRICANIZED HONEY BEES ARE A MIX OF AFRICAN BEES AND EUROPEAN BEES. THEY WENT TO SOUTH AMERICA AND SPREAD TO OTHER PLACES, INCLUDING PARTS OF AFRICA.

INVASIVE SPECIES

INDIAN HOUSE CROWS, HAVE SPREAD TO VARIOUS
PARTS OF THE WORLD, INCLUDING EASTERN
AFRICA, SINGAPORE, AND PARTS OF EUROPE,
LARGELY DUE TO BEING TRANSPORTED BY SHIPS.

INVASIVE SPECIES

FERAL CATS ARE CATS THAT USED TO LIVE WITH PEOPLE BUT NOW LIVE WILD OUTSIDE. THEY CAN BE FOUND IN MANY PLACES, INCLUDING NORTH AMERICAN GRASSLANDS, WHERE THEY DON'T BELONG.

ILLEGAL HUNTING

ELEPHANTS HAVE BIG TUSKS MADE OF A HARD, WHITE MATERIAL CALLED IVORY. SOME PEOPLE WANT IVORY TO MAKE JEWELRY, STATUES, OR DECORATIONS.

ILLEGAL HUNTING

PANGOLINS ARE SHY, SCALY ANIMALS THAT LIVE IN ASIA AND AFRICA. THEY USE THEIR HARD SCALES TO PROTECT THEMSELVES FROM DANGER.

ILLEGAL HUNTING

RHINOS ARE BIG ANIMALS WITH HORNS ON THEIR NOSES. THEY USE THEM TO STAY SAFE AND FIND FOOD.

ILLEGAL HUNTING

HAWKS ARE BIRDS OF PREY THAT HUNT FOR FOOD, LIKE SMALL ANIMALS AND BIRDS. THEY ARE IMPORTANT FOR KEEPING NATURE IN BALANCE.

ILLEGAL HUNTING

DEER LIVE IN FORESTS AND GRASSLANDS.
PEOPLE HUNT DEER FOR THEIR MEAT AND
ANTLERS, BUT SOMETIMES IT IS ILLEGAL.

ILLEGAL HUNTING

LEOPARDS LIVE IN GRASSLANDS AND HAVE
BEAUTIFUL SPOTTED FUR. SOME PEOPLE
ILLEGALLY HUNT THEM TO TAKE THEIR SKINS
AND SELL THEM FOR MONEY.

ILLEGAL HUNTING

PRAIRIE DOGS ARE SMALL ANIMALS THAT DIG
HOLES AND LIVE IN GROUPS. THEY HELP THE
GRASSLANDS AND ARE FOOD FOR OTHER ANIMALS.

OVERGRAZING

FERAL GOATS ARE WILD GOATS IN AUSTRALIA. THEY EAT A LOT OF GRASS AND PLANTS. IF THERE ARE TOO MANY, THEY CAN MAKE THE SOIL DRY AND UNHEALTHY, WHICH LEADS TO OVERGRAZING.

OVERGRAZING

WILDEBEESTS ARE HERBIVORES THAT GRAZE ON GRASS IN THE AFRICAN GRASSLANDS. WHILE WILDEBEESTS ARE IMPORTANT TO THE ECOSYSTEM, IF THERE ARE TOO MANY OF THEM IN ONE AREA, THEY CAN OVERGRAZE THE GRASS.

"THE PRAIRIE HAS A SUBTLE BEAUTY THAT REVEALS ITSELF SLOWLY TO THOSE WHO WATCH AND LISTEN."

-INSPIRED BY EARLY 19TH-CENTURY NATURALIST WRITINGS

NOTES

www.ingramcontent.com/pod-product-compliance
Lightning Source LLC
Chambersburg PA
CBHW080256150626

46556CB00024B/3343